Spring Math Walk

Deanna Pecaski McLennan

For Quinn who finds joy in nature

Copyright © by Deanna Pecaski McLennan
First edition 2020

All rights reserved.

No part of this publication may be reproduced in any form, or by any means, electronic or mechanical, including photocopying, recording, or any information browsing, storage or retrieval system, without permission in writing from the author.

www.mrsmclennan.blogspot.ca

Joyful
Math

The sun warms me as I walk through the fields. Spring is a time of growth and change. If I look closely I notice math all around me.

A turtle walks along my path.

Its geometric shell looks like armour.

How old might he be?

Nestled in the grass I find a feather.

I feel its delicate edges.

Why does it have a pattern?

In the soil something is growing.

Circular leaves are beginning to open.

How did this plant survive the winter?

Up in the tree I hear an opossum.

It looks at me before scurrying away.

Why is she still awake?

I gently touch a peony in the garden.

It is a perfect sphere.

How do so many petals fit inside?

Farther ahead I find an old tree.

A maze is carved on its bark.

What creature has tunnelled here?

A brood of mallards waddle along.

The ducklings keep close to their mother.

Why do they have webbed feet?

I hear something rustling in the stones.

I detect a camouflaged snake.

How does it move without legs?

I smell the fragrant sand cherry shrub.

The dainty flowers are so pretty.

Why are there five petals on each?

A dandelion pushes through the grass.

Its happy face makes me smile.

Why does it open at the same time each day?

I gather a bunch of tulips.

I leave their bulbs in the soil.

Why do the flowers look identical?

The tree buds are ready to open.

Green leaves will unfold from within.

How do they know when it is time?

A friend joins me to splash in a puddle.

Our boots sink into the mud below.

How deep is the water?

We hear a robin's call from above.

A mother is resting in the tree.

How did she build such a round nest?

My brother picks a dandelion.

He blows the seeds and makes a wish.

How far will they travel?

After my walk I play in the yard.

The grass smells so fresh and sweet.

How fast can I roll?

Author's Note

Spending time immersed in nature is a wonderful way for young children to learn about our world.

There is beauty in mathematics. Helping children to recognize the strong connection between nature and math may encourage them to see mathematics as an aesthetic and captivating subject. So often the only experiences children have with math are those found inside the classroom. Exploring the authentic math that exists in nature may help nurture children's interest and confidence, building a strong foundation for subsequent experiences.

My hope in writing this book is to inspire children, educators and families to see math as an inviting discipline that lives beyond the walls of the classroom. Our natural world is filled with amazing mathematical connections. This book does not need to be read beginning to end. The photos can be used individually, or in combination, to spark mathematical conversations and connections with children. Ask children what they see, think and wonder about each picture. Ask what their theories are for what they see happening on each page. At the end of the book you will find information to complement each photo. Adults can support and extend children's mathematical and scientific ideas using this information. Additional resources can scaffold and build inquiries that spark from the text.

The information presented in this book can serve as an introduction to new math concepts, or as a reference when natural treasures are discovered by children outdoors. Consider reading it together with children before venturing out into the world on your own math walk. You might choose to use the photos as conversation starters, or read the book in its entirety using photos and narrative.

When we look at the world through a mathematical lens, we discover that anything is possible!

~Deanna

In this photo draw children's attention to the turtle's large shell and tail. Ask children to describe the shapes and designs they spot on the shell, and hypothesize what advantages there might be to having such a spiky tail.

Snapping turtles live in fresh water including ponds and ditches. They can live past 100 years of age, and hibernate in the winter by burying their bodies in mud at the bottom of lakes and rivers. They are slow moving and cannot retreat into their shells for protection. They use their strong jaws and flexible necks to defend against enemies.

Math ideas might include measurement, speed, density, numbers, strength, shapes and patterns.

In this photo draw children's attention to the characteristics of the feather, focusing on colour, shape, and size. Ask children to hypothesize from which bird the feathers belong, and how they might have travelled to the spot on the grass.

Birds have feathers for many reasons - to attract mates, to provide protection from injury, to aid with flying, and for insulation against the elements. We can identify birds by their feathers' specific characteristics. Feathers also help birds camouflage in order to avoid detection by predators.

Math ideas might include classification, sorting, measurement, spatial reasoning, geometry, patterns, numbers, and data management.

In this photo draw children's attention to the delicate manner in which the stonecrop buds are gently opening. Ask children to describe what they notice about the plant (e.g., its shape, size, density), and hypothesize what it will look like when it grows to full size.

Stonecrops are hearty, flower-bearing perennials that are drought resistant because they are capable of storing water in their leaves. They tolerate cold temperatures well and can be grown directly in the ground or in pots. Their delicate flowers bloom in late summer and early fall and are a great source of food for pollinators including bees, bats, beetles and butterflies.

Math ideas might include measurement, temperature, size, lifecycles, shape, colour, and patterns.

In this photo draw children's attention to the opossum hiding in the tree. Ask children to hypothesize whether it is a male or female, and describe how they think it climbed so high.

Male opossums are larger than females and have yellow fur on their chests. Opossums are marsupials who live in woodlands, farmlands, and residential areas. They are typically nocturnal and not very social, preferring to live solitary lives close to food and water sources. They enjoy scavenging through trash and eat meat and vegetation including insects, worms, and snakes. When they feel threatened they play dead, staying still for up to six hours in order to trick their predators.

Math ideas might include time, colour, measurement, classification and sorting.

In this photo draw children's attention to the thick peony bloom at the top of the slender stem. Ask children to hypothesize why the bloom is shaped like a sphere, how long it might take for the flower to fully open, and what it might look like when all petals are exposed.

Peonies are fluffy, fragrant flowers that are popular in residential gardens. They come in a number of colours and grow quite tall. Ants are attracted to a peony's nectar, and help it open when they climb inside the bud. If you look closely at the photo and you will see one! They are perennials that live for more than one growing season without needing to be replanted. Some peony bushes can grow for over one hundred years!

Math ideas might include shape, size, colour, classification, data management, numbers, patterns, and seasons.

In this photo draw children's attention to the intricate design that appears on the tree's bark. Ask children to hypothesize how and why the markings were created. Challenge them to find the starting point of the markings, and identify as many different lines as they can (e.g., loops, straight, zigzag).

Emerald Ash Borers are non-native beetles and an invasive species to the Great Lakes provinces and states. It is believed that these beetles arrived in North America from Asia by attaching themselves to wood packing that was used in cargo ships or airplanes. These pests attack Ash trees when their larvae tunnel through the tree's vascular system, resulting in the tree's death. This affects the local habitat, with a loss of shelter and food for other living things dependent on the tree.

Math ideas might include shape, size, lifecycle, colour, classification, numbers, measurement, direction and problem solving.

In this photo draw children's attention to the manner in which the ducklings are walking together with their mother. Ask children to hypothesize what they think the ducks might be doing, and how the mother communicates with her babies. A group of ducklings is called a 'brood'. See what other names children might know for collections of animals (e.g., flock, herd).

Ducklings recognize their mother and siblings through a phenomenon know as 'imprinting' - an impression made in their brain shortly after birth. Mallard ducks bring their ducklings to water the day after they are born. The faster a duckling learns to swim and forage food on its own, the better its chance of survival. A duck's webbed feet act as paddles and help it swim more efficiently in the water.

Math ideas might include similarities and differences, aerodynamics, classification, sorting, print making, reflection, and numbers.

In this photo draw children's attention to the snake camouflaged in the stones. Ask children to describe the advantages of camouflage for an animal. Ask children to hypothesize how the snake travelled to this place without legs. In the photo it appears as though the snake has captured something. If appropriate for the age of the readers, ask children to discuss the predatory nature of snakes and how they eat large pieces of food.

Snakes use patterned muscle contractions to help them move. Different muscles contract or relax at synchronized times, resulting in sidewinding and wavelike movements. Snakes can also eat objects larger than they are. They have flexible skulls and jaws, which hinge in different places, allowing them to open their mouths very wide.

Math ideas might include patterns, direction, locomotion, size, shape, length, colour, angles and cycles.

In this photo draw children's attention to the delicate sand cherry flowers and rich purple foliage of the shrub. Ask children to estimate how many flowers are in the picture, and problem solve how many petals there are in total (e.g., using the anchor of five since each flower has five petals).

Sand cherry plants are perennials that blossom in early spring. They can be pruned into different shapes and sizes. The branches are self-similar, containing patterns. This means the whole has the same shape as many of its smaller components. A small section of this shrub looks like the entire tree, just in miniature. The branches are fractals. These patterns repeat smaller and smaller copies of themselves over and over.

Math ideas might include fractals, patterns, self-similarity, anchor numbers, colour, shape, size, number, estimation, and measurement.

In this photo draw children's attention to the small dandelion flower pushing through the grass. Ask children to describe what they see, and hypothesize how the dandelion bud is able to push through the hard ground and dried grass and still blossom.

In addition to being one of the first sources of pollen for insects in the spring, dandelion leaves are a source of nutrition for humans as well. They are edible and rich with vitamins, minerals and fiber. Some people eat dandelion greens in salads, or drink them in tea. Dandelion flowers tend to regularly open in the morning and close at night. This predictable behaviour is called nyctinasty, and is a circadian rhythm in plants. It is thought to help a flower protect itself from nighttime elements including cold temperatures and moisture.

Math ideas might include time, temperature, shape, density, size, nutrition, sense, cycles, and patterns.

In this photo draw children's attention to the shape, colour and size of the tulips. Ask children to describe what they see and hypothesize what the tulips might look like when they are in full bloom.

Tulips grow from bulbs that are planted in the fall before the ground freezes, and usually begin to grow during early spring in March. Tulips are available in a variety of colours, and are cup-shaped, with one flower growing on each individual stem. These beautiful flowers are almost perfectly symmetrical. They have three petals and three sepals, making it appear as though they have six petals. Many animals including deer, mice, voles, and squirrels love to eat tulip bulbs directly from the garden.

Math ideas might include colour, shape, size, numbers, symmetry, doubling, addition, temperature, and seasons.

In this photo draw children's attention to the shape and size of the leaf bud. Ask children to hypothesize how the leaves will emerge and form. Ask children to hypothesize how buds know when it is the right time of year to open.

In the winter trees are dormant. They sense the cold through 'chill hours' and instinctively know to stay asleep. They slowly wake up as the days get warmer and there is more sun during the daytime (because the nights grow shorter). Sometimes trees can be tricked into thinking it is spring during unseasonably warmer temperatures in the winter months. This can cause frost damage to their buds, affecting fruit production later on in the warmer months.

Math ideas might include temperature, light, shadow, patterns, lifecycles, and calendar.

In this photo draw children's attention to how much water is in the large puddle. Ask children to describe how they might go about measuring the depth of the water, or the capacity of water contained in the puddle.

A puddle usually forms where there is a depression in the surface where the water is located. In this photo the puddle formed on a patch of dirt resulting in a thick layer of mud at the bottom. When light reflects off the smooth surface of water, an image is produced. Because the water had a slight ripple to it, the reflection of the children's boots appears wavy in this photo. The light is being reflected in different directions. This is called diffuse reflection.

Math ideas might include measurement, reflection, doubling, capacity, cycles and temperature.

In this photo draw children's attention to the intricately formed robin's nest. Ask children to describe what they see, and hypothesize how the bird built such a perfectly round nest.

In order to build a nest a robin collects about 350 dried pieces of grass and twigs. After a rain, when the ground becomes wet, a robin collects hundreds of beakfuls of mud. The robin uses her beak to weave the grass and twigs, and uses the mud as a type of cement to secure it all together. The mud also ensures that the nest is securely affixed to a support like a tree branch or window ledge. In order to create the round shape, the robin pushes her breast in and around the inside of the nest. This helps shape it into the right size for the babies and mother to fit inside. The nest keeps the babies warm and dry.

Math ideas might include shape, size, problem solving, circumference, measurement and numbers.

In this photo draw children's attention to the dried dandelion seeds that the boy is blowing from the seed head. Ask children to make connections between the picture and their own lives. Ask children to consider why each individual seed is shaped the way it is, and how far the seeds might travel in the wind after being blown.

Each dandelion seed has approximately 100 'pappus', which act as tiny parachutes. When horizontal winds blow the pappus help the seed travel as far from the dandelion as possible. This aids the dandelion to reproduce and generate many plants. Some seeds can travel up to a kilometre away from their original location.

Math ideas might include lifecycles, shape, size, distance, aerodynamics, and direction.

Deanna Pecaski McLennan, Ph.D., is an elementary educator in Ontario, Canada. Deanna is fascinated by math and loves exploring its natural and authentic application in the living world. She hopes to help children and families recognize math as a beautiful and fascinating subject, and grow children's confidence, accuracy and interest in math.

Follow Deanna on Twitter and Instagram for regular updates including ideas for engaging children in playful, emergent math inside the classroom and beyond. Extending math learning outdoors is a favourite exploration!

Connect with Deanna:

deannapecaskimclennan@gmail.com
@McLennan1977

Also from Deanna

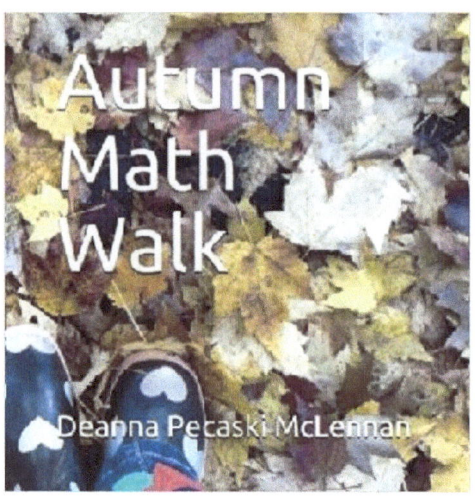
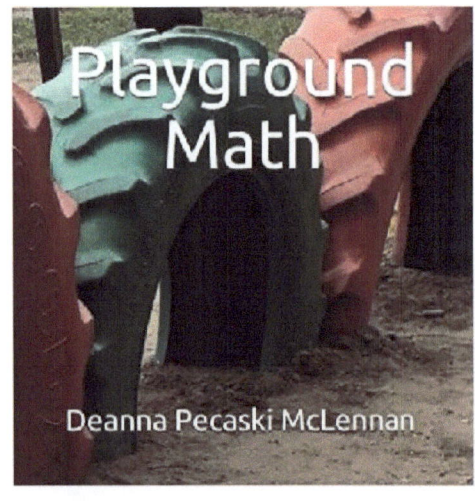

Joyful
Math

www.ingramcontent.com/pod-product-compliance
Lightning Source LLC
Chambersburg PA
CBHW051215220526
45473CB00003B/1039